By Dr. Ref, PhD
Scitenberg Kids

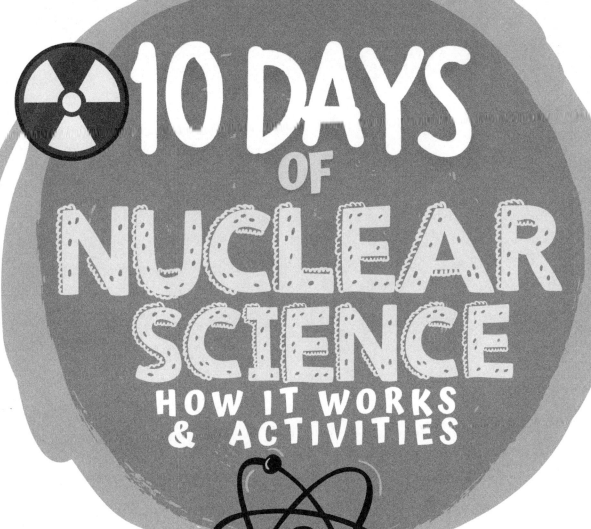

10 DAYS
OF
NUCLEAR
SCIENCE
HOW IT WORKS
& ACTIVITIES

RADIOACTIVITY, NUCLEAR FISSION, NUCLEAR MEDICINE, AND MORE!

10 Days of Nuclear Science

THE ATOM

What is an Atom made of?

Each atom has a nucleus. Inside the atom are three particles: the proton and neutrons in the nucleus and the electrons that orbit the nucleus. Atoms are grouped into different families in the periodic table of elements.

X Mass number: Number of protons + Number of neutrons

Z Atomic number: number of protons

$$_Z^X A$$

Symbol of the atom

Nucleus

Proton

Neutron

Electron

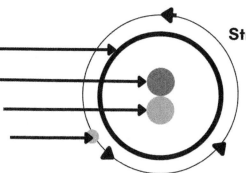

Structure of an atom

How it all started . . .

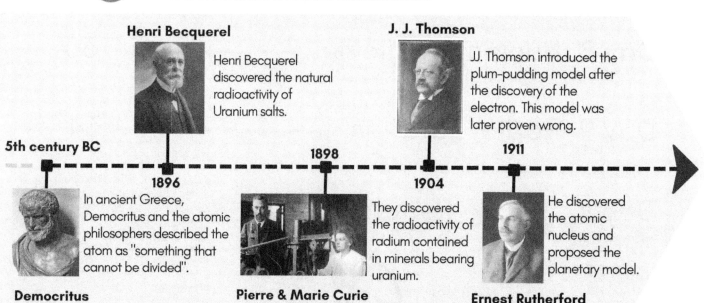

Henri Becquerel

Henri Becquerel discovered the natural radioactivity of Uranium salts.

J. J. Thomson

JJ. Thomson introduced the plum-pudding model after the discovery of the electron. This model was later proven wrong.

5th century BC

1896

In ancient Greece, Democritus and the atomic philosophers described the atom as "something that cannot be divided".

Democritus

1898

1904

They discovered the radioactivity of radium contained in minerals bearing uranium.

Pierre & Marie Curie

1911

He discovered the atomic nucleus and proposed the planetary model.

Ernest Rutherford

THE ATOM

Marie Curie (1867-1934) was the first woman to win a Nobel Prize in 1903 for the discovery of natural radioactivity, along with Henri Becquerel and Pierre Curie. She went on to win another Nobel Prize, making her the only woman to win two Nobel Prizes in the history of science. Her second prize was for the discovery of radium in 1911.

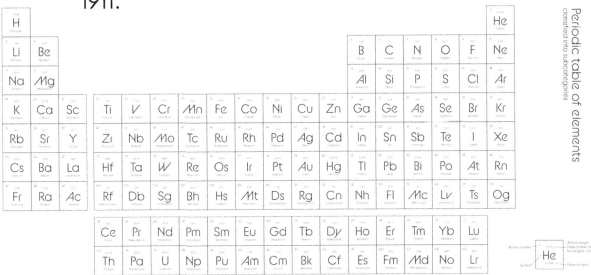

The Periodic Table of Elements

Each element is made up of atoms.

Atoms of the same element have a specific number of proton and their number of electrons is equal to the number of protons.

The same element can have a different number of neutrons.

An element that has the same number of protons but a different number of neutrons is called an **isotope**.

THE ATOM

Test your knowledge

Read the sentences below and write T for true, and F for false.

An element can have a different number of neutrons. ———

Electrons are located in the nucleus. ———

Marie Curie received 2 Nobel prizes. ———

Atoms have the same number of protons and electrons. _____

Name three physicists.

- _____

- _____

- _____

Name the three subparticles in an atom.

- _____

- _____

- _____

THE ATOM

Test your knowledge

The study of atoms has never stopped, complete the timeline below.

Ask an adult to help you choose a website that you can use for your research.

Robert Oppenheimer - Niels Bohr - Enrico Fermi - Irene and Frederic Joliot-Curie

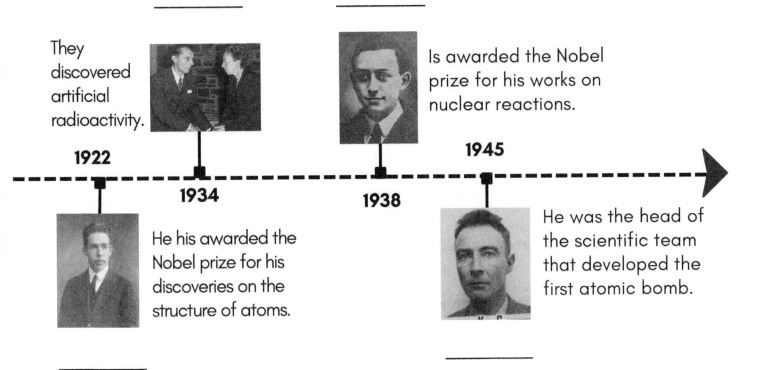

They discovered artificial radioactivity.

1922

1934

He his awarded the Nobel prize for his discoveries on the structure of atoms.

Is awarded the Nobel prize for his works on nuclear reactions.

1945

1938

He was the head of the scientific team that developed the first atomic bomb.

RADIOACTIVITY

Some atoms are unstable. They rearrange their nucleus to try to become more stable by ejecting particles. This phenomenon is called radioactivity.

Which atoms are radioactive?

Very heavy atoms reorganise their structure to become lighter. There are different types of radioactivity. Alpha, beta positive, beta negative, gamma and more.

α Radioactivity

The atom releases an alpha particle (Helium atom). It loses 2 protons, 2 electrons and 2 neutrons.

$$^{240}_{94}\text{Pu} \longrightarrow \, ^{236}_{92}\text{U} \; + \; ^{4}_{2}\text{He}$$

Plutonium   Uranium   Helium

β⁻ Radioactivity

The atom releases an electon and gains a proton.

$$^{14}_{6}\text{C} \longrightarrow \, ^{14}_{7}\text{N} \; + \; ^{0}_{-1}\text{e}^{-}$$

Carbon   Nitrogen   electron

RADIOACTIVITY

γ Radioactivity

The atom is excited, it releases energy under the form of gamma rays.

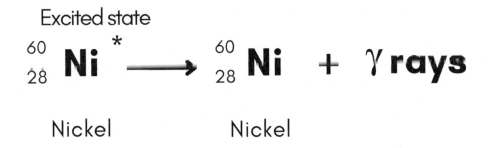

Excited state

$$^{60}_{28}\text{Ni}^* \longrightarrow {}^{60}_{28}\text{Ni} + \gamma \text{ rays}$$

Nickel   Nickel

What is an excited state?

An atom is in an excited state when its energy is higher than that of the ground state.

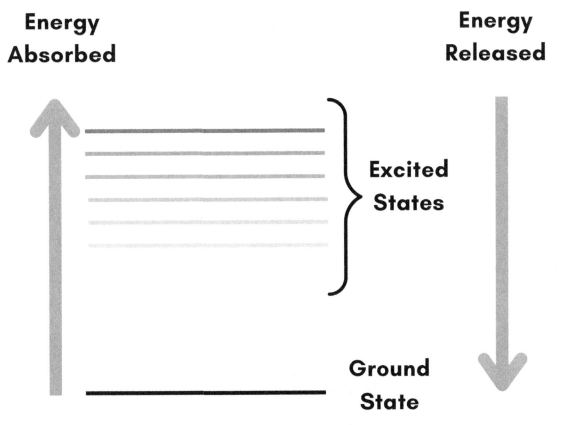

Energy Absorbed

Energy Released

Excited States

Ground State

RADIOACTIVITY

Test your knowledge

Complete the following equations and fill in the blanks.

$$^{---}_{94}\text{Pu} \longrightarrow ^{236}_{92}\text{U} + ^{-}_{2}\text{He}$$

Plutonium   Uranium   Helium

$$^{14}_{6}\text{C} \longrightarrow ^{14}_{7}\text{N} + ^{0}_{-1}\text{e}^-$$

Carbon   _ _ _ _   _ _ _ _

$$^{60}_{28}\text{Ni}^* \longrightarrow ^{60}_{28}\text{Ni} + \text{\_\_\_\_}$$

_ _ _ _ _ _   _ _ _ _ _ _

_ _ _ _ _ _   _ _ _ _ _ _

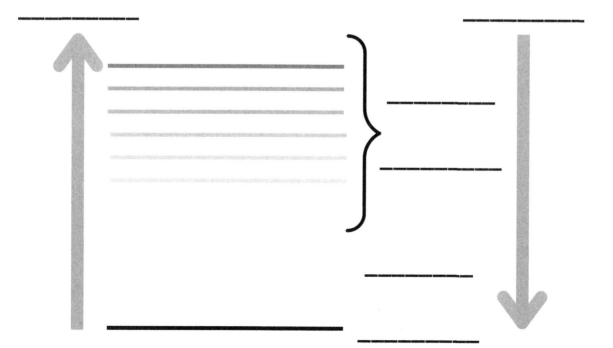

RADIOACTIVITY

Test your knowledge

Matching activity.
Match each statement to the type of radioactivity it belongs.

Involves the loss of 1 Helium atom. •

Involves the loss of an electron. •

• **α Radioactivity**

The resulting atom has a higher mass than the original atom. •

• **β⁻ Radioactivity**

The resulting atom has a lower mass than the original atom. •

NUCLEAR FISSION

Nuclear fission occurs when a neutron hits a heavy nucleus. This nucleus splits into two lighter atoms and releases about 2 to 3 neutrons.

What are the applications of nuclear fission?

Nuclear fission reactions release a lot of energy. They can be used to produce electricity and atomic bombs.

neutron   heavy nucleus   lighter nuclei

In nuclear reactors, it is the fission of Uranium-236 that is used to produce energy. First, Uranium-235 captures a neutron to become Uranium 236.

$$_{0}^{1}\text{n} + _{92}^{235}\text{U} \longrightarrow _{92}^{236}\text{U}$$

Uranium   Uranium

Uranium 236 spontaneously fissions into two fission products, Barium and Krypton, and releases 2 to 3 neutrons.

$$_{92}^{236}\text{U} \longrightarrow _{56}^{141}\text{Ba} + _{36}^{92}\text{Kr} + 3\,_{0}^{1}\text{n}$$

Uranium   Barium   Krypton

NUCLEAR FISSION

The mass of uranium-235 is greater than the mass of the added barium and krypton.

The heavy nucleus is heavier than the two fission products because there are strong nuclear interactions in heavy atoms. The difference in mass can be used to produce energy.

Did you know that?

Each nuclear fission reaction releases about two hundred million eV (200 MeV) of energy.

0.7% of uranium is uranium 235 which can fission, and 99.3% of the remaining uranium is uranium 238 which cannot fission.

1 gram of uranium that fully fissions produces as much energy as 2.8 tonnes of coal or 1.6 tonnes of oil.

2 million years ago, in the Oklo uranium mines in Gabon, a nuclear fission reaction took place over several hundred thousand years. The water that flooded the mine slowed down the neutrons from the sun, which then initiated the fission reaction.

NUCLEAR FUSION

Nuclear fusion occurs when two light nuclei fuse to form a heavier nucleus.
This type of reaction occurs in the core of stars.

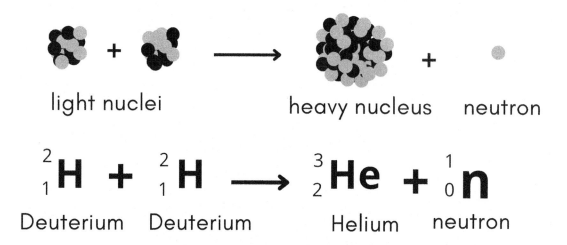

light nuclei   heavy nucleus   neutron

$$^{2}_{1}H + ^{2}_{1}H \longrightarrow ^{3}_{2}He + ^{1}_{0}n$$

Deuterium Deuterium   Helium   neutron

These reactions can take place due to the extreme temperature (almost 16 million degrees) and pressure conditions in the core of the stars.

The two light nuclei have the same charge (positive) and therefore tend to repel each other.

These reactions only take place in the stars. To reproduce them on Earth, the two light atoms must be attracted to each other. There are two different methods for achieving this by using a **tokamak** or a **stellarator**.

NUCLEAR FUSION

With both techniques, light atoms are attracted to each other under very extreme conditions. The elements are confined in a plasma (very hot gases) with a very high magnetic field.

So far, the longest fusion reaction on earth took place during 5 seconds. It produced 59 megajoules of heat, which corresponds to 60 kettles boiling.

Why nuclear fusion?

The energy released by the fusion of two light atoms is higher than the energy released by the fission of heavy atoms. Nuclear fusion releases 4 times more energy than fission, 4 million times more energy than burning oil, coal or gas.

NUCLEAR FUSION

 Test your knowledge

Read the sentences below and fill in the blanks with the appropriate words.

Nuclear fusion occurs when two _____ nuclei give a bigger atom. It releases _____ energy than nuclear fission.

Nuclear fusion occurs in the core of _____ . It is not yet possible to use it on earth yet due to the extreme _____ and _____ conditions needed for the reaction to occur.

Read the sentences below and write T for true, and F for false.

Nuclear fusion can be done with big atoms. _____

Nuclear fusion occurs in the sun. _____

Fusion usually leads to the formation of an helium atom. _____

During a nuclear fusion reaction, particle are ejected. _____

Nuclear fusion releases more energy than burning coal. _____

NUCLEAR FUSION

 Test your knowledge

Complete the nuclear reactions below with the missing atom, particle, atomic number or mass number.

$$_\_^2H + {}_\_^2H \longrightarrow {}_2^\_He + {}_0^1n$$

Deuterium   Deuterium   Helium   _____

$$_\_^{236}U \longrightarrow {}_{56}^\_Ba + {}_{36}^{92}Kr + 3{}_0^1n$$

_____   Barium   Krypton   _____

$$_0^1n + {}_\_^{235}U \longrightarrow {}_{92}^\_U$$

_____   _____   _____

Try to find out how much raw material is needed to produce the same amount of energy.

Uranium: 1g   Coal: _

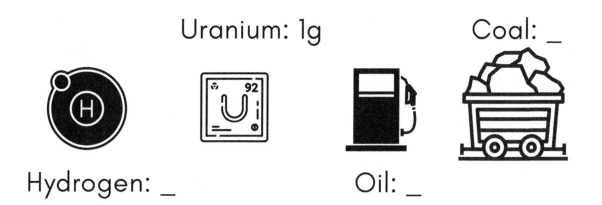

Hydrogen: _   Oil: _

NUCLEAR POWER PLANTS

Nuclear power plants use the energy released by the fission of uranium atoms to produce electricity.

When a uranium atom fissions, three neutrons are released. Each neutron can then collide with another uranium atom and generate another fission.
This is called a **chain reaction**.

A nuclear power plant has several components:

The core of the reactor

This is where nuclear reactions take place. In most of today's reactors, uranium fission is used. The energy released is used to boil water, water also being a moderator (it slows down the neutrons).

The turbine

The steam passes through the turbine and makes it turn. It converts heat into mechanical energy.

The generator

The generator converts mechanical energy into electricity.
The generator is connected to transmission lines that supply the houses.

NUCLEAR POWER PLANTS

The steam is then sent to cooling towers, where it is evacuated. The steam is not radioactive, the cooling and steam generation system are separated from the reactor core.

Did you know that?

The production of electricity in nuclear power plants does not produce carbon dioxide.

If an accident occurs, the release of radioactivity can be very dangerous. This is why the reactor is separated from the outside world by at least three layers of protection.

In most reactors, the pressure of the core is higher than ambient pressure.

These reactors are called pressurised water reactors.

There are currently 450 nuclear reactors operating worldwide.

Nuclear energy is the second source of low carbon electricity worldwide behind hydro-electricity.

NUCLEAR POWER PLANTS

 Test your knowledge

Read the sentences below and fill in the blanks with appropriate words from the list below.

reactor   uranium   generator fission steam turbine
transmission lines carbon dioxyde

A   nuclear   _____   uses   a   _____   to   turn mechanical energy into electricity.

The   _____   _____   linked   to   the   generator power houses.

_____   reactions   of   _____   atoms   are   used   to create heat.

The water is turned into   _____   due to the heat in a nuclear   _____.

Electricity generation in nuclear power plants does not release any   _____   _____.

The   _____   is used to turn the heat released by _____ reactions into mechanical energy.

NUCLEAR POWER PLANTS

 Test your knowledge

Unscramble the words below and write the answers in the boxes on the right.

rminua	
meats	
cearotr	
rtonegare	
neirutb	
reco	
issonif	
ecatoinr	
omta	
neryge	

CHAIN REACTIONS

For a chain reaction to take place, the first neutron must collide with a nucleus, releasing other neutrons which in turn collide with nuclei. Under certain conditions, the reaction is self-perpetuating.

How to control a chain reaction?

To control a chain reaction, the number of neutrons escaping from the nuclei must be controlled. Certain elements such as water or boron will absorb the neutrons and allow the reaction to be controlled.

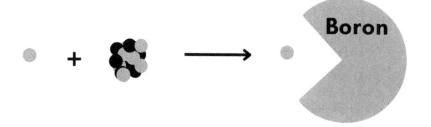

What happens if the chain reaction is not controlled?

If the chain reaction is not controlled, all the energy can be released at once, resulting in a huge explosion.

CHAIN REACTIONS

In a chain reaction, the fission of each atom releases neutrons that can participate in fission reactions.

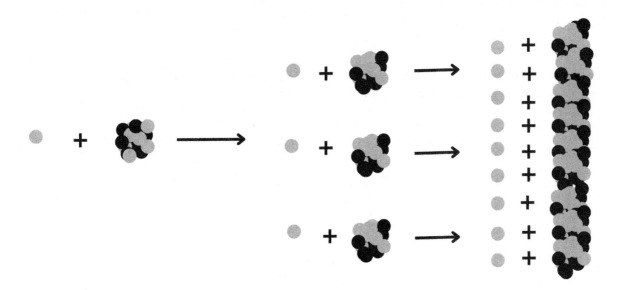

In a nuclear reactor, the chain reaction is controlled: water and boron absorb the neutrons to avoid an out-of-control situation.

In a nuclear bomb, the chain reaction is not controlled, the energy is released in a single event.

How much neutrons are released?

In nuclear reactors, an average of 3 neutrons are released from the fission of uranium-236. The most common elements able to give a chain reaction are uranium and plutonium.

CHAIN REACTIONS

 Test your knowledge

Unscramble the following words to find the names of the most common atoms giving chain reactions.

IMUAUNR

– – – – – – –

IMULUNPOT

– – – – – – – – –

How many neutrons are liberated by the fission of an uranium atom?

–

Which chemicals can be used to absorb neutrons?

– – – – –

CHAIN REACTIONS

 Test your knowledge

Complete the scheme below with the missing neutrons and atoms to draw a chain reaction.

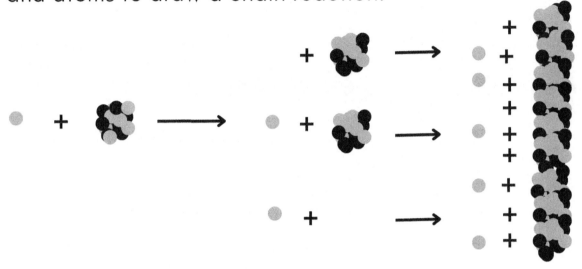

How many neutrons are liberated by the fission of 4 uranium atoms?

_ _

If 9 neutrons are liberated, how many uranium atoms have fissioned?

_

If 15 neutrons are liberated, how many uranium atoms have fissioned?

_ _

NUCLEAR WASTE

Nuclear waste is waste containing radioactive materials. Radioactive waste is produced by several types of activities, such as medicine, research, energy production, extraction and purification of certain minerals...

How is nuclear waste produced?

Nuclear waste is produced mainly in nuclear reactors. The heavy atom splits into two pieces. These two pieces cannot fission and are radioactive. There are also undesirable nuclear reactions where atoms capture neutrons instead of fissioning, which creates new nuclei and does not produce electricity.

Why is that a problem?

Radioactivity is very dangerous for humans and the environment. Some radioactive elements are dangerous for very long periods (several billion years).

Nuclear waste is disposed of deep underground in specially created areas, where it will remain for centuries.

NUCLEAR WASTE

There are three kind of nuclear waste

Low level waste

Radioactive for less
than a 100 years

Intermediate level waste

Radioactive for up
to a billion year

High level waste  Radioactive for several
billions of years

Did you know that?

Each radioactive atom has a half-life. After each half-life,
half of the atoms disappear. After a period of 10 half-lives,
less than 0.1% of the original atoms are still present.

Plutonium has a half-life of 24,500 years.
It takes 245,000 years to see more than 99% of the
plutonium disappear.
There are deep geological storage sites in operation or in
the process of being built in Finland, Sweden, USA, France.

NUCLEAR WASTE

 Test your knowledge

Read the sentences below and fill in the blanks with appropriate words.

Nuclear waste is waste that contains _____ materials.

The half-life period corresponds to the time needed for half of the radioactive _____ to naturally disappear.

After 10 half-lives, __ % of the initial atoms are still present.

Radioactivity is very dangerous for _____ and the _____ .

Complete the following scheme.

___ **level waste**

Radioactive for less than a ___ years

_____ **level waste**

Radioactive for up to a _____ year

____ **level waste** Radioactive for several _____ of years

NUCLEAR WASTE

Search your favorite website for interesting facts about nuclear waste and nuclear waste management.

Write them down below.

_____

_____

_____

_____

_____

_____

_____

_____

NUCLEAR MEDICINE

Nuclear medicine works by using radioactive atoms to treat diseases or to obtain images of the damaged body.

Here are two ways in which nuclear medicine is used:

Diagnosis

Nuclear medicine is used to obtain very precise images of human tissue. Radioactive elements can be used, such as technetium.

Treatment

Radioactivity can 'burn' tissue, like a sunburn. Radioactive elements are used to treat damaged organs and cells.

What is a PET scan?

Positron-emission tomography is used to produce image of the inside of the body. A radioactive molecule is injected in the body. A camera will then detect radiation and, by contrast, find areas where the body is not working as it should.

NUCLEAR MEDICINE

What is a radiotherapy?

A radiotherapy is used to destroy cells that have a disease. By colliding particles on a damaged cell or using the radioactivity of a radioactive metal absorbed by the patient. There are 2 types of radiotherapies: external and internal radiotherapy.

External radiotherapy

Radiation source

Radiation beam

Damaged cell: target

Internal radiotherapy

Radioactive liquid
absorbed by the patient

Damaged cell: target

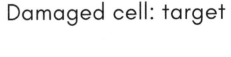

Emitted
particle

NUCLEAR MEDICINE

 Test your knowledge

Name two types of radiotherapies.

_____

_____

Which radiotherapy involves the absorption of a radioactive element by the patient?

_____

Why do we use radioactive elements in hospitals?

_____

Can you name a medical analysis based on radioactivity?

_____

Can you name a radioactive metal used in hospitals?

_____

NUCLEAR MEDICINE

DAY 8

 Test your knowledge

In the box below, draw and label a scheme showing the principle of external radiotherapy.

Label the scheme below explaining an internal radiotherapy:

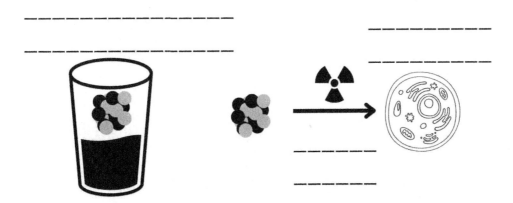

RADIO-DATATION

The number of atoms in a radioactive element decreases with time. This property can be used to obtain the age of very old samples.

Carbon-14 is the best known dating method for obtaining the age of archaeological objects.

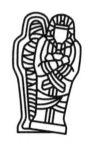

The tree is alive, it exchanges carbon with the atmosphere.

The tree dies, it stops exchanging carbon

The wood is used to make objects.

*Every **5,730 years**, half of the carbon atoms are turned into nitrogen 14.*

Initially

... 5,730 years later

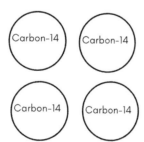

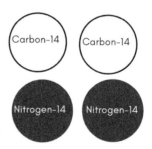

I am 5,730 years old!

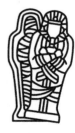

By counting the carbon and nitrogen atoms in the sample, the age of the specimen can be estimated.

RADIO-DATATION

Carbon-14 dating can be used for samples up to **60,000 years** old. For older samples, there is not enough carbon left.

Carbon-14 dating is made difficult for samples younger than **200 years**. Two hundred years ago, humans began releasing large amounts of carbon into the atmosphere as a result of the industrial revolution. This changed the natural concentration of carbon in the atmosphere.

Since the 1940s, man has detonated hundreds of atomic bombs, releasing artificial radioactive elements into the atmosphere. This has also made dating methods less accurate.

Uranium/lead dating is also a very famous method to date minerals

Uranium naturally turns into lead over time.

By estimating the lead/uranium ratio with scientific equipment, the age of samples older than one billion years can be estimated. This method can only be applied to minerals containing uranium.

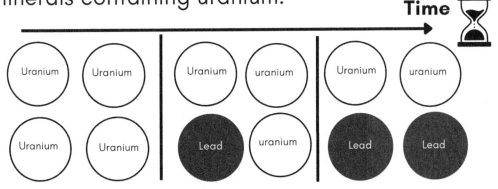

RADIO-DATATION

 Test your knowledge

Fill in the blanks with appropriate words.

Carbon 14 can be used to date objects up to _____ years old.

Every _____ years, half of the carbon 14 atoms dissapear.

Carbon 14 is turned into _____ 14.

To date very old minerals, uranium- _____ datation can be used. It can be used for minerals more than a _____ years old.

Why is carbon 14 datation becoming less precise recently?

_____

Can we use carbon 14 datation to determine the age of a living tree?

_____

RADIO-DATATION

Complete the scheme below.

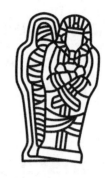

The tree is alive, it _____ with the atmosphere.

The tree dies, it _____ exchanging

The wood is used to make objects.

Initially

After _____ years

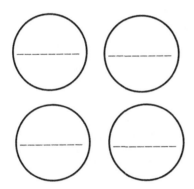

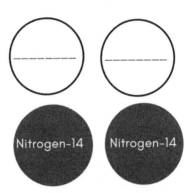

10,000 carbon atoms were found in the wood of a dying tree.

How many carbon 14 atoms will remain in 5,730 years?

_____

How many carbon 14 atoms will remain in 57,300 years?

_____

FUTURE REACTORS

Current nuclear reactors produce nuclear waste. Future nuclear reactors will be able to use nuclear waste as fuel and will be safer. They may even produce their own fuel.

What is fast neutrons?

Fast neutrons are not slowed down by water. By using liquid metals such as sodium as a cooling medium, uranium-238 will capture neutrons to produce energy.

How to make a machine that produce its own fuel?

The uranium-238 will capture neutrons and create fissile material. It is possible to create more fissile material than the initial amount of atoms, and thus 'create' fuel.

$$_{0}^{1}n + _{92}^{238}U \longrightarrow _{92}^{239}U \longrightarrow _{93}^{239}Np \longrightarrow _{94}^{239}Pu$$

FUTURE REACTORS

Why do thorium reactors are interesting?

Thorium is an abundant element on earth. Future thorium reactors are considered safer. If the temperature is increasing, the nuclear reactions become less probable and the reactor cools down.

Thorium captures a neutron, give a fissile element, fission occurs and releases energy

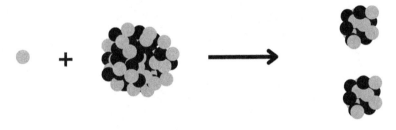

More fission, temperature increases

As temperature increases, it is more difficult for neutrons to reach their target

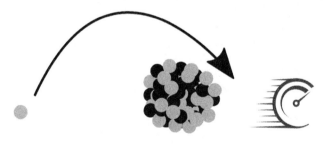

Less fission, temperature decreases

In case of issues, an emergency tank placed under the reactor recovers the radioactive materials.

FUTURE REACTORS

Test your knowledge

Watch a documentary about nuclear reactors and draw a scheme showing a nuclear reactor with the following elements: reactor vessel, cooling system, generator, cooling tower, turbine, containment building.

FUTURE REACTORS

 Test your knowledge

Unscramble the words below and write the answers in the boxes on the left.

hotiurm	
fsta	
ntoreun	
dosimu	
melente	
flisise	
claoont	
nleucra wtsae	
lufe	
petinunmu	

What new words have you learned?

Please feel free to evaluate and leave a review.

We'd love to hear from you so we can create quality content for you.

Follow our page for updates on new releases and improved recommendations.

Thank you!

Printed in Great Britain
by Amazon